Dr. Frank A. Forte

Unveiling The World Of Single-Domain Protein Catenanes

From Design To Future Applications

This book was professionally typeset on Reedsy

Find out more at reedsy.com

Contents

1.

2.

3.

4.

5.

6.

7.

8.

9.

10.

Introduction

"Unveiling the World of Single-Domain Protein Catenanes: From Design to Future Applications" is an in-depth look at a novel family of proteins defined by two mechanically linked polypeptide rings that fold synergistically into a compact and integrated structure. The book discusses single-domain protein catenanes' design, synthesis, characterization, functional features, applications, and future prospects. Single-domain protein catenanes are designed and synthesized by rewiring the connectivity between secondary motifs, adding artificial entanglement, and producing the protein in cells through a sequence of controlled, streamlined post-translational processing steps. The book sheds light on the possible applications and consequences of these unique protein molecules in a variety of sectors, providing vital insights into a game-changing innovation in protein engineering. The book is a must-read for protein engineering, biomaterials, industrial enzymes, and drug discovery researchers, students, and professionals.

Understanding Protein Catenanes With A Single Domain

Single-domain protein catenanes are a new protein class distinguished by two mechanically linked polypeptide rings that fold synergistically into a compact and integrated structure. This one-of-a-kind design is created by rewiring the connectivity between secondary motifs, creating artificial entanglement, and producing the protein in cells via a sequence of controlled, streamlined post-translational processing processes with no further in vitro reactions.

Several proteins, notably dihydrofolate reductase (DHFR) and green fluorescent protein (GFP), have been used to synthesize single-domain protein catenanes. The DHFR catenane, known as cat-DHFR, has been fully described and found to have improved stability and resilience. When compared to its linear cousin, it has better anti-aggregation capabilities and a greater thermal melting point (Tm). Despite a loss in catalytic activity, cat-DHFR has higher heat resistance, keeping more than 70% of its catalytic activity following a 10-minute incubation at 70 °C, whereas the linear control loses practically all activity.

The presence of leucine, an amino acid found in high concentrations in whey protein, is linked to muscle protein synthesis. Whey protein, which is high in leucine, can deliver the amino acids needed to enhance muscle protein synthesis and minimize muscle protein breakdown during exercise.

Single-domain protein catenanes represent a substantial improvement in protein engineering, providing improved stability, robustness, and prospective applications in a variety of sectors. Incorporating leucine-rich whey protein into a pre-workout meal helps to enhance muscle protein synthesis while decreasing muscle protein breakdown after exercise.

Background Information About Protein Catenanes

Background information on protein catenanes includes their discovery in natural systems, cellular synthesis, and prospective applications. Protein catenanes are two mechanically connected polypeptide rings that fold synergistically into a compact and integrated structure, a design obtained by rewiring secondary motif linkage. They have been discovered in natural systems, such as the X-ray crystal structure of the bacteriophage HK97s capsid, as well as synthesized in cellular environments, as evidenced by unambiguous proof of catenane topology via techniques such as SDS-PAGE, SEC, and partial digestion experiments. Single-domain protein catenanes, such as the cat-DHFR and green fluorescent protein catenanes, have been designed and synthesized, displaying higher stability, anti-aggregation characteristics, and increased temperature robustness. These protein catenanes have a wide range of applications, including industrial enzymes, antibodies, cytokines, and biomaterials. The synthesis of protein catenanes has opened up new avenues in chemistry and provides a scaffold for the creation of fusion protein catenanes, allowing for the investigation of the impact of their topology on structure-property interactions. The availability of these single-domain protein catenanes allows for the investigation of topological impacts on structure-property interactions and opens up new ground for protein molecules, transcending the linear paradigm of natural protein molecules.

Single-domain Protein Catenanes: Design And Synthesis

The design and synthesis of single-domain protein catenanes entail rewiring the connectivity between secondary motifs to introduce artificial entanglement, resulting in the formation of two mechanically interlocked polypeptide rings that fold synergistically into a compact and integrated structure. This novel technique has been applied successfully to a variety of proteins, including dihydrofolate reductase (DHFR) and green fluorescent protein (GFP).

The following are important parts of the design and synthesis process:

1. Rewiring connectivity: The design entails rewiring connectivity between distinct secondary motifs in order to introduce endogenous entanglement, which results in the production of single-domain protein catenanes.

2. Using model proteins: Model proteins such as GFP have been used to build single-domain protein catenanes, indicating the method's adaptability to other proteins with similar folds.

3. Synthesis: Single-domain protein catenanes can be synthesized in two phases via a pseudorotaxane intermediate or directly in cells using planned, streamlined post-translational processing activities.

4. Fusion protein catenanes: The method enables the production of fusion protein catenanes by introducing proteins of interest at loop areas, resulting in thermal and mechanical stability due to strong conformational coupling.

The design and synthesis of single-domain protein catenanes expands chemistry and provides a strong scaffold for a variety of applications, including industrial enzymes, antibodies, cytokines, and   biomaterials.

Single-Domain Protein Catenanes Characterization

The characterization of single-domain protein catenanes, such as cat-DHFR and cat-GFP, has received a lot of attention. These catenanes, which are made up of two mechanically interconnected polypeptide rings that fold together to form a compact and integrated structure, have shown improved stability, robustness, and unique functional features.

Various experimental techniques were used to characterize single-domain protein catenanes, including combined SDS-PAGE, size-exclusion chromatography (SEC), liquid chromatography-mass spectrometry (LC-MS), ion mobility spectrometry-mass spectrometry (IMS-MS), and proteolytic digestion experiments. These findings conclusively demonstrated the topology of cat-DHFR and cat-GFP, revealing structural and functional features.

The findings indicate that single-domain protein catenanes, such as cat-DHFR, have better anti-aggregation capabilities and stronger heat resistance than their linear counterparts. The cat-DHFR, for example, has a higher thermal melting point (Tm) and retains more than 70% of its catalytic activity after a 10-minute incubation at 70 °C, whereas the linear control loses practically all activity.

Single-domain protein catenanes' characterization has cleared the road for their prospective uses in a variety of sectors, including industrial enzymes, antibodies, cytokines, and biomaterials. Furthermore, the availability of these catenanes has permitted the understanding of topological impacts on structure-property correlations, providing a new paradigm for protein design and engineering.

The characterization of single-domain protein catenanes has provided vital insights into their unique structural and functional features, opening up new avenues for application and research.

Single-Domain Protein Catenanes Functional Properties

The functional properties of single-domain protein catenanes, such as cat-DHFR and cat-GFP, have been studied, indicating improved stability, activity, and unique traits. Cellular synthesis of protein catenanes has been shown to improve the stability and activity of folded structural domains, opening up a new paradigm for protein design and engineering.

Cat-DHFR, the single-domain protein catenane of DHFR, has been comprehensively described and shown to have improved anti-aggregation capabilities and higher thermal resistance than its linear cousin. Despite a decrease in catalytic activity, cat-DHFR retains a considerable amount of its activity even after being exposed to high temperatures, making it an attractive option for a variety of applications, such as industrial enzymes and biomaterials.

Similarly, the single-domain green fluorescent protein catenane, cat-GFP, has been developed, synthesized, and characterized, indicating the possibility of creating fusion protein catenanes with improved thermal and mechanical stability. This novel method provides a diverse technique for the design and synthesis of single-domain protein catenanes, hence opening up new avenues in chemistry and protein engineering.

The functional properties of single-domain protein catenanes, such as improved stability, activity, and the ability to construct fusion protein catenanes, hold great promise for a wide range of applications, representing a significant advancement in the field of protein engineering and design.

Single-Domain Protein Catenanes In Use

Single-domain protein catenanes, such as cat-DHFR and cat-GFP, have a wide range of applications and offer great potential in a variety of domains. The synthesis of single-domain protein catenanes from linear protein precursors with well-preserved functionalities has opened up new avenues of research in chemistry and protein engineering.

Some of the most important applications and implications of single-domain protein catenanes are as follows:

1. **Improved stability and resistance:** When compared to their linear counterparts, single-domain protein catenanes have been demonstrated to have improved stability, anti-aggregation capabilities, and higher thermal resistance. As a result, they are intriguing candidates for use in industrial enzymes, biomaterials, and other fields where stability is critical.

2. **Fusion Protein Potential:** The design and synthesis of single-domain protein catenanes opens the door to the development of fusion protein catenanes. Because of the tight conformational connection, this technique provides thermal and mechanical stability, bringing new opportunities for the

construction of novel protein molecules with better
functional capabilities.

3. **Topological Effects on Structure-Property Connections:**
   The availability of single-domain protein catenanes has aided
   in the understanding of topological effects on structure-
   property connections. This opens up a new paradigm for
   protein molecule design and engineering, with implications
   for understanding and modifying protein function.

Single-domain protein catenanes have numerous uses, including
improved stability, the potential for fusion proteins, and the
investigation of topological implications on structure-property
connections. These unique protein molecules have a high potential
for use in a variety of practical applications, and they represent a
substantial achievement in the field of protein engineering and
design.

Effects Of Topology On Structure-property Relationships

The implications of topology on structure-property interactions have been studied in a variety of domains, including materials science, chemistry, and protein engineering. The study of the properties of objects that are preserved throughout continuous deformations such as stretching, bending, or twisting without tearing or gluing is referred to as topology.

Topology is important for the unique structural and functional features of single-domain protein catenanes. As two polypeptide rings interlock in catenanes, they form a topologically complex structure with increased stability, robustness, and unique functional capabilities as compared to their linear counterparts.

The availability of single-domain protein catenanes has aided in the understanding of topological implications for structure-property correlations, opening up a new paradigm for protein design and engineering. The topological effects on protein structure and function have consequences for understanding and regulating protein function, with potential applications in a wide range of domains such as industrial enzymes, biomaterials, and drug development.

Topological impacts on structure-property interactions are an important component of single-domain protein catenanes, providing a new paradigm for protein design and engineering. Topological effects have implications for understanding and modifying protein function and have potential applications in a variety of domains.

Mapping The Protein Universe Into Single-domain Protein Catenanes

The mapping of the protein universe into single-domain protein catenanes has been studied, and the results have been promising. As two polypeptide rings interlock in catenanes, they form a topologically complex structure with increased stability, robustness, and unique functional capabilities as compared to their linear counterparts. The design and synthesis of single-domain protein catenanes provide a diverse technique for the building of fusion protein catenanes with improved thermal and mechanical stability, allowing for the design of novel protein molecules with improved functional capabilities.

A recent study has shown that the existing linear protein universe can be mapped into single-domain protein catenanes with well-preserved functionalities. A single-domain catenated dihydrofolate reductase (cat-DHFR) and a single-domain green fluorescent protein catenane (cat-GFP) have been designed and synthesized, illustrating the method's flexibility for additional proteins with comparable folds. Single-domain protein catenanes can be synthesized in two processes: via a pseudorotaxane intermediate or directly in cells by planned, streamlined post-translational processing mechanisms. The availability of single-domain protein catenanes has aided in the understanding of topological implications for structure-property correlations, opening up a new paradigm for protein design and engineering.

Mapping the protein universe into single-domain protein catenanes represents a substantial improvement in protein engineering and design, providing a diverse technique for the construction of fusion protein catenanes with improved functional features. The availability of single-domain protein catenanes has aided in the understanding of topological implications for structure-property interactions, opening up new avenues for protein design and engineering.

Prospects And Implications For The Future

Single-domain protein catenanes have important future prospects and consequences, with possible applications in a variety of fields. While the search results do not directly address the future prospects and implications of single-domain protein catenanes, we can build on existing field information.

Future prospects

1. **Biotechnological Applications:** Single-domain protein catenanes have the potential to be used in a variety of biotechnological applications, such as the production of innovative industrial enzymes, biomaterials, and drug delivery systems.

2. **Protein Engineering:** The design and synthesis of single-domain protein catenanes represents a new paradigm for protein engineering, with the potential to construct custom-designed proteins with improved stability and functional capabilities.

Implications:

1. Therapeutic Applications: The improved stability and distinct functional features of single-domain protein catenanes may have significance for the development of novel protein-based therapies.

2. Biophysical experiments: The availability of single-domain protein catenanes allows for biophysical experiments to be conducted to better understand the impact of protein topology on protein structure and function.

Single-domain protein catenanes' future possibilities include possible uses in biotechnology and protein engineering, as well as therapeutic applications and biophysical studies. Additional prospects and implications for these unique protein molecules are anticipated to be discovered through further research in this field.

Conclusion

The book "Unveiling the World of Single-Domain Protein Catenanes: From Design to Future Applications" examines the design, synthesis, characterization, functional properties, applications, and future prospects of single-domain protein catenanes in depth. This book provides unique insights into a game-changing breakthrough in protein engineering, shedding light on the possible applications and consequences of these novel protein molecules in a variety of fields.